Cosmogony:
or thoughts on philosophy

JOHN MERRILL

compiled, edited, and augmented by
the author's 3rd great grandson,
Brian Lee Merrill

TABLE OF CONTENTS

JOHN MERRILL, the "Philosopher of the Pool"

Excerpts from the "Milwaukee Journal" of Sunday, September 23rd, 1928:

> In the annals of the little town of Pardeeville, picturesquely situated on the Fox River and two lakes, John Merrill, a settler from New Hampshire, who was related to S.S. Merrill, an early resident of Milwaukee, and to Henry Merrill, sutler at Old Fort Winnebago, occupies a conspicuous niche.

In John Merrill's life were many unique incidents. He had a fair education, was widely read of an ingenious turn of mind, with a bent for natural science. While living in New Hampshire he wrote a book entitled "Cosmogony or Thoughts on Philosophy" which contained a refutation of Isaac Newton's theory of gravitation. Mr. Merrill contended that the center of the earth consists only of space and that our globe consists is composed of various layers, earth, air, and water in regular order. He held that at each pole there is a large hole, and that into the one at the North Pole the ships of missing Arctic explorers sank. He described the agricultural activities of the inhabitants of the interior of our planet and the way in which the sun's rays reached the region. It will be remembered that Bulwer Lytton wrote a novel called the "Coming Race" in which he pictured a race that lived underground.

Mr. Merrill's manuscript was polished and copied with pen and ink by a woman resident of Pardeeville and printed in pamphlet form in New Hampshire. Many copies were sold in those early days especially in his native state, where he owned a celebrated nook in the White Mountains called the Pool. It was situated at Franconia Notch, at the foot of Mount Lafayette, in a canyon on the

Pemigewasset River. Each summer after settling in Pardeeville he made a trip to his pool, where from tourists he gathered a harvest of dollars for the use of his boats.

Naturally John Merrill did a flourishing business in his books, as well as his boats. He gave lectures on his theory before many famous persons, including a number from abroad, and he wrote a letter to Queen Victoria and sent her a copy of his scientific work.

One eventful day he received what *purported* to be a reply from the ruler of the British Empire. A copy was made by a printer in New Hampshire and is in the possession of Charles W Merrill of Pardeeville, John Merrill's grandson. It is yellowed by age. The original is supposed to have been lost or inadvertently destroyed after the death of John Merrill at the home of his daughter in Pardeeville.

In 1888 an eastern newspaper under the caption "The Philosopher of the Pool" printed a complimentary mention of John Merrill with the following letter from him:

> *Mr. Editor: Please say to my friends that I have retired from The Pool, after being there 34 years, and concluded to spend the rest of my days on the homestead in Pardeeville, Wis. Where I can sit and*

see the 100 acres of crops almost ready for harvest. Crops never looked better. Give my best and respects to all, till we meet in heaven. I am almost home. This is my eighty-seventh year. Am well, only old age says stay with my children.

Yours in love,
JOHN MERRILL
From Pardeeville

Thus after having paddles his skiff so many years, the old philosopher drops his oars with this plaintive strain, and thus too, a rugged landmark, second only in importance to his more aged rival, the "Old Man of the Mountains," disappears from view. The author philosopher died in 1892 at the age of 90 years and was buried in the Pardeeville cemetery. On his monument is inscribed a map of our globe, on which he spent so much time and thought. He is remembered as one of the picturesque figures in Wisconsin history.

COSMOGONY: OR THOUGHTS ON PHILOSOPHY

FOREWORD
By Brian Lee Merrill

Many times, it is more about the journey than the destination. People, in general, are inclined to accept the world view of the "learned experts," and by extension all the conclusions they propose, by whatever methods they choose to employ, without pursuing it themselves. A few people of the whole, might end up at the same destination as those experts, but take a completely different path to get there, on the operations of their own thought, and by different methods of their choosing ... most definitely to their credit.

And fewer still take their own journey and end up at a different destination entirely. No matter how likely or unlikely the conclusion, employing our own brand of reason and logic, as the Lord Heavenly Father intended, is the journey we should all endeavor to take. To paraphrase Robert Frost, "I took the path less traveled by, and that has made all the difference."

John Merrill definitely took the path less travelled with his take on Hollow Earth theory. He lays it out with observation from his own surroundings, and that from the reports of Arctic explorers of his day. Though his theory could be evaluated with ridicule and speculation 150 years ago, and most certainly could be so today, some parts have yet to be conclusively disproven.

John Merrill comes to the conclusion that the thickness of the earth's shell is eight miles, "from outside air to inside air," as he calls it. As of the year 2014, the deepest hole ever drilled into the earth is reported to be about 7.6 miles. According to Merrill, this would place the drill only less than a half mile from breaking through on the inside surface of the earth. Had the drilling project continued with success, Merrill's assumptions may have been proven correct, surely to the shock of the vast majority – or alternatively, at least, his assumptions of thickness would have been proven inaccurate, or at most, his theory proven completely in error. But alas, eight miles in depth has not been reached, so the physical "last word" has yet to be spoken on the issue to this day, over 150 years later.

As for John Merrill's suspected polar entrances to the interior, even today, the exact construct of the Polar region as well as Antarctica, it can be argued, is not conclusively known beyond all doubt. Admiral Bird's "Operation Highjump" in 1948 is popularly characterized as the final word of dismissal on Hollow Earth Theory, but there are several circumstances, rumors, and (many would say) fringe conspiracies as to the actual outcome of Operation Highjump - enough for many minds to question whether Hollow Earth is truly dismissed, and for others to conclude that Hollow Earth is indeed purposely concealed fact.

In addition, Merrill also did not subscribe to Newton's theory of gravity, given his own views on those natural forces. Since his time, we have had many brilliant minds weigh in on gravity, of course to include Albert Einstein. With all the science we know in relation, we today tend to regard the theory of gravity as fact, but it is still a theory, and has not yet ascended to the level of scientific law.

Now granted, I believe the extensions of Hollow Earth that connect Ancient Alien Theory and/or time travel are probably far beyond John Merrill's own intended scope, but I feel he invited the journey of thought, no matter the direction, to see where the journey might take us. And in this objective I believe he succeeded, as the letters *"he received"* indicate - even as they were likely only satire from his 19th century contemporaries, his theory provoked enough thought and debate to cause their creation - because he took the path less travelled.

PREFACE

The author of these pages begs that a critical public will not be too hasty in expressing an opinion for or against the theories therein advanced. For if they be correct it will require more than a cursory glance at the arguments to enable one to decide their merits.

And if they are fallacious, he feels assured it will require some thought to refute them.

He has endeavored to be as concise as possible and has made his work as short as convenient and clearly convey his ideas. Should it meet with a good degree of success, it may be followed by a larger and more respectable work.

It is now cast upon the waters of public opinion trusting that seed will fall into the minds of some and bring forth good fruit in due season.

CREATION

A dark, chaotic, formless deep,
Did earth appear at first,
And doomed to an eternal sleep,
Till into life it burst.

God spake, - and said "Let there be light,"
And soon the rolling sphere
Formed sun, moon, and stars of night,
Dry land, and seas appear.

Yes, Nature robed in varied hue,
Obeyed the great behest;
Creation's countless millions flew
To climes that suit them best.

But still the work was not complete,
God said, "Let us make man,
And take of him and make a meet,
To cheer and help him plan.

"We'll stamp our image on his brow
And set the seal of love,
And bid him rule his subjects now
As we do those above."

D. A.

COSMOGONY

The formation of the earth has always been a subject of deep interest to men of science. Much has been said and much has been written in regard to its construction, its shape and the laws governing it. For a long period it was supposed to be flat and exceedingly limited in its extent, bounded on all sides by oceans the limits of which no one pretended to know. What are now among the most enlightened portions of the world were mentioned by Herodotus as belonging, at that time, to the extreme ends of the earth. Millions have lived and died the existence of whom the wisest sages of Greece and Rome did not know, or even suspect. Nor had they any knowledge of the spherical form of the earth, or its annual or diurnal motions. And when this great fact was, hundreds of years later, first promulgated, scarcely a believer could be found, and those who taught the belief were only considered worthy the insane asylum or the darkest dungeon. And so, too, when Columbus declared to the eastern continent the existence of a new world beyond the waters of the Atlantic, the supposition was considered ridiculous, and for years one of the world's greatest benefactors wandered about, the jest of his unbelieving countrymen.

Similar treatment has been the lot of nearly every new theorist in treating of the earth or other planets. An intelligent world should learn from the past not to treat new theories with contempt, at least not until a fair investigation be given them, lest they fall into the same errors that have so often blinded others. Let them remember that though a theory may appear absurd and ridiculous to them, it may still be correct. The principles they now advocate with such tenacity were once considered as ludicrous as these new innovations on their philosophy.

By my own observations, with what I have been able to glean from the works of eminent travelers, it appears to me that much of the natural philosophy as now taught is incorrect. I am also convinced that the generally received opinions in regard to the formation of this world and its habitable parts are no less correct.

That there are yet undiscovered continents teeming with animal and vegetable life, the inhabitants of which may enjoy all the advantages that we enjoy, seems to me more than probable. The evidence is abundant and clear that this earth is not a solid sphere, but a hollow world, more flattened at the extremities than is usually admitted; that it is open at the northern and southern extremities admitting heat, light, air and space inside; that there are continents and oceans within as habitable and navigable as those on the outside.

In the beginning when the elements of space, air, earth and water were one chaotic mass, "without form and void," the Creator put forth His hand as described in

Genesis, bidding each element take its appointed position. Then the element of earth being more solid than the others appeared first; then the element of water appeared being more solid than those remaining; the air then came forth covering the whole face of the earth, and lastly yhe element of space surrounding everything made. And certain laws were given to each element, earth, water, air, and space. At this time also, the other heavenly bodies were created, each composed of the same elements as our own, and they took their appropriate places in the great frame-work of infinite space, according to their magnitude or weight, influenced however, by a diversity of incidental causes.

The element of space before spoken of, it will be noticed, is that which lies above or beyond the element of air, surrounding it in every position. It is much finer and lighter than air, as air is finer and lighter than water, and water lighter than earth. And as a ball of equal weight as a given amount of water ten or more feet below the surface, will sink that distance and there remain; or as a balloon, lighter than the atmosphere on the surface of the earth, will rise till it reaches a point where the air is of the same weight, where it also rests, so will a planet sink into space until its weight is all absorbed and will remain there subject more or less, to the laws of attraction from neighboring planets or more powerful magnetic bodies inducing its several revolutions.

It is found that a body weighing a given amount of the earth's surface will weigh less as it approaches the higher or more rarefied air. Now if the same bodies could be weighed again after reaching a certain distance in our element of space they would be found to lose their weight entirely. In like manner a planet sinking into this same element loses its weight and continues to abide there, revolving as first put in motion by the Creator.

It may be well to remark in this connection that the principle of weight as usually taught is, in my view, incorrect. It is said that when Sir Isaac Newton saw the apple fall to the ground instead of rising into the air he attributed it, after mature deliberation, to the power of the earth to attract all substances, a power technically called "attraction of gravitation." The force with which this power acts upon different bodies is called "weight."

Here is a fallacy I would correct by showing that weight is simply a reaction of the element in which the substance dwells. Thus a piece of light wood weighing ten pounds in the element of air is not acted upon with sufficient force to destroy its entire weight; but place it in a more solid and heavier element as water, and it loses its weight altogether, owing to the greater reaction of this element.

This may be more forcibly illustrated in the following manner. It will be remembered, however, in the first place that I contend this world is a hollow, nearly spherical, body. The ocean has been sounded in some one hundred places and found to be about four miles deep. In some places, as in the Gulf Stream and in some salt as well as fresh water lakes it has been found impossible to find a bottom at all. Now we will suppose there are four miles of earth to air inside, the continents and oceans on the inside and on the outside being opposite each other. This will give us a distance of eight miles from air outside to air inside.

Now we will suppose we have a cylindrical tube in the shape of a horse shoe magnet, four miles from the top of each end to the curve; this will give from one end round to the other, eight miles, just equal to our hypothetical distance from inside to outside air. Let this tube be first filled with air, and a leaden ball be dropped into each end; they will both seek a resting place at the bottom. Let two balloons be placed at the bottom and one will seek the top of each end. So all bodies heavier than air will seek the bottom and which is the earth; and all bodies lighter than the air will rise until they are equal with the air, or till their weight is absorbed by it, or the counter-reaction, so to speak, balances them, and here they will remain.

So also, all waters in the air sufficiently condensed to be heavier than the air, and all waters on the earth seek a level by pressure, or reaction of a lighter element, and all that cannot displace the particles of air in the top of the earth and take its place will run off till a level is found.

Let us suppose again that the tube be filled with water. Take two leaden balls as before and they will again seek the bottom till their weight is absorbed. Then take two corks, or balloons as before, both of which will seek the top, the latter with greater velocity than the corks, and also with greater force than in the element of air. So all bodies dwelling in this element, heavier than the element itself, will seek the bottom till its weight is all absorbed; and all bodies lighter than the element itself will rise till the same result is produced.

Next we will fill the tube with dry sand; this being dry will contain more or less of the element of air. Now if we pour water upon the sand it will press down, displacing the air and taking its place. Again we fill the tube with sand, let water be admitted at the bottom. Then, water being an element lighter than sand it seeks the top of both tubes, where it meets the element of air which is lighter than itself, and it can go no further.

Now if we pour mercury into both tubes, it will expel the water and take its place, the same as water expels air, and if sufficient be poured in it will drive out both water and sand. Here we see again that a lighter element in a heavier one will seek a level or equilibrium toward an element of similar density; and that a heavier element in a lighter one will seek an equilibrium toward a heavier one, or at the bottom of the element less dense. This explains the philosophy of gasses rising and water falling; and why liquids in the earth lighter than itself, seek the top.

My idea of weight is further illustrated by the atmosphere. We take one pint of water from its element and weigh it in the lighter element of air and it will weigh, we say, one pound: allowing fifteen pounds of atmospheric pressure to every square inch, we get 2160 pounds of pressure to every square foot on the earth's surface, which would make a column of water a square foot in size thirty-two feet high. If this column were in a tube and a hole made in the bottom, the force with which the water presses though this hole will be equal to the weight of the air on the top of the water; for shut off the air from the top of this column, and a child will place his finger upon the hole and prevent its pressing out, and if the air was perfectly removed from its top it would not press out at all. Cutting off the air from this water destroys its weight entirely, the same as if it were again put into its own element. The same result is here produced as would be if the column were in the element of space till its weight was destroyed; which is virtually the case in this experiment.

The hot winds peculiar to tropical regions furnish another illustration of the truth of our position. The following extract will show the effect of these winds.

> "Hot winds are very frequent in countries contiguous to the tropical regions. Large deserts and plains, covered with little vegetation, engender very warm winds; these winds, which are

of a noxious character, prevail in the vast deserts of Asia and Africa, where they show themselves in all their force. Nubia, Arabia, Persia, and other parts of Asia, are visited by a burning wind particular to the desert. In Arabia it is called samoun, from the Arabic samma, which signifies hot and poisonous. It is also named samiel, from samm, posion. In Egypt it is called chamsin (fifty) because it blows for fifty days, from the end of April until June, at the commencement of the inundation of the Nile. In the western part of the Sahara it is named harmatten.

"The Simoon is announced by the troubled appearance of the horizon; afterward the sky becomes obscured, and the sun loses its brilliancy, - paler than the moon, its light no longer projects a shadow; the green of the trees appears of a dirty blue, the birds are restless, and the affrighted animals wander in all directions. The rapid evaporation occurring at the surface of the human body dries the skin, inflames the throat, accelerates respiration, and causes a violent thirst. The water contained in the skin evaporates, and the caravan is a prey to all the horrors of thirst.

"It usually lasts three days, but if it exceeds that time it becomes insupportable. Woe to the traveler whom this wind surprises remote from shelter! He must suffer all its dreadful consequences, which sometimes are mortal. The danger is most imminent when it blows in squalls, for then the rapidity of the wind increases the heat to such a degree as to cause sudden death. This death is a real suffocation; the lungs, being empty, are convulsed; the circulation disordered, and the whole mass of blood, driven by the heat toward the head and breast; whence that hemorrhage at the nose and mouth which happens after death.

"This wind is especially fatal to persons of a plethoric habit, and those in whom fatigue has destroyed the tone of the muscles and vessels. The corpse remains a long time warm, swells, turns blue, and is easily separated; all of which are signs of that putrid fermentation which takes place when the humors become stagnant. These accidents are to be avoided by stopping the nose and mouth with handkerchiefs; an efficacious method is also that practiced by the camels, who bury their noses in the sand, and keep them there till the squall is over."

The destructive influences of these winds are owing to the rarefaction of the atmosphere which destroys the weight of all bodies according to the degree of its heat. In this case the intense heat of the earth at these times so rarefies the air that bodies such as sand are found in higher clouds, and is also taken up with logs, trees and other heavy substances and carried about with ease. The effects on the human body are similar to those produced on ascending high mountains when respiration becomes difficult, and violent thirst is produced. Similar results but in a much less degree are witnessed in the hot seasons, even in our own latitude. It is here not uncommon to hear people complaining that it is "so hot they can hardly breathe." The same principle explains why we drink more water in the summer season than in winter.

Again, iron when melted loses much of its weight because its great heat rarefies the air pressing upon it, and it would not press out of the hole supposed to be made in the bottom of the above tube, with so much force as the water, though it is much heavier when cold.

COSMOGONY: OR THOUGHTS ON PHILOSOPHY

It is now presumed that we are fully understood in regard to our idea of the principle of weight, and we now proceed to treat, of the opening at the north and south extremities letting air and space together with heat and light to the inhabitants supposed to dwell therein. Although we may have no positive evidence of our theory, yet we have evidence that will go far to support it, and we think convince every unprejudiced mind that the probabilities are in our favor.

The reader's attention is now called to the following extract from Dr. Kane's first voyage page 298. –

> "Feb. 9. To-day we had a sky of serene purity, and all hands went out for a sanitary game of romps in the cold light. Presently three suns came to greet us – strange Arctic parhelia – and a great golden cross of yellow brightness, uniting them in one system. Under the glare of them we played foot-ball! At meridian we made a rough horizon of the ice, and found ourselves in latitude about 72° 16'. At this time another marvel rose before us – land! The monster was to the W.S.W. in the shape of two, round-topped hills, lifted up for the time in our field of view. An hour or two later, while the day was waning, these hills became mountains, and then a line of truncated cones, the spectre of some distant coast. Looking out a few moments later, while the day was waning – out of the little door of our felt house, to where for the last fortnight a bleak sameness of snow has been stretching to the far north, we saw a couple of icebergs standing alone in the sky!"

This scenery is produced by the sun's refractions, as was also the following illustration given the author by the venerable Thomas Barnard Esq., for a long period a resident on his farm in Orange, N.H., where the phenomenon was seen. Many may have observed similar occurrences before. He has often told me that when on a particular place near his residence, which is situated on the height of land between Merrimack and Connecticut Rivers, and when the sun is in a particular place in the west, just above the Green Mountains, he can distinctly see Mascomy Lake, situated in Enfield, though at all other times it is invisible. Also when standing on the adjacent Mount Cardigan toward evening, the whole range of the Green Mountains is visible in the heavens if there be a cloud overhanging them; and to the east, in the morning Lake Winnipiseogee is also visible.

This is in all cases, owing to the refraction of light, by which the rays are brought to the line of vision.

Dr. Kane in describing the Aurora Borealis invariably tells us that it is always seen in the form of a circle or half-circle. It is produced by the sun's reflection, upon clouds emerging from the extremity. For Kane's remarks on the Aurora see the following extract from the thirty-fifth chapter of his work commencing on page 314.

> "Their changes seemed to be dependent upon modifications rather of intensity than form. They were characterized by neither active movement nor varied coloring. My tabular observations will be published elsewhere, but I subjoin a rude attempt at analysis of their distinctive features.

1st. A mere illumination, apparently emerging from a dark cloud some five degrees above the horizon, more resembling a nebulous patch or a moonlight cirrus than the auroral light.

2nd. Detached bands of illumination, impressed against the sky, like a condensed nebulosity, unconnected with any visible central arc, and distributed near about the line of the magnetic axis between the horizon and the zenith. These were sometimes stratiform, converging by perspective, and reminding one of the auroral plates, *plaques aurorales* of Lottin.

3rd. A well-marked zone or band, or sometimes several concentric ones, either broken or continuous, unaccompanied by the ordinary segments of light or cloud, passing through or near the zenith in a direction which, according to the mean of some fourteen observations, was sixteen degrees east of the magnetic meridian. These bands were constantly varying, not by active scintillation, but by changes of intensity – rapid flashing augmentation, sudden subsidence, or complete extinction – a wavy oscillation, resembling wind action.

4th. Bistre-colored clouds, assuming a segmentary or arch-like form, and throwing out rays of clear, white light; these streaming up toward the zenith, and sometimes across to the opposite horizon, with more of coruscating movement than any other form. It was somewhat remarkable, that of six such displays observed in October and February, every one was in the direction of the sun, then not more than eight degrees below the horizon, and in one instance above it – a true daylight aurora. These jets, although not colored, might be looked upon as rudimentary forms of those dependent rays, now recognized by observers as corresponding in direction with the local magnetic inclination.

If we regard these forms as characterizing generally the auroras of this region, we can not help being struck with their departure from the indications observed by Lieut. Hood, in the Franklin Expedition of 1820. His observations may be referred to two general classes. The first commencing with arches, either to the east or west of the magnetic meridian, or coincident with it, sometimes four or five in number, rising in concentric series, and never less than 5° in altitude: these, upon reaching the zenith, become broad, irregular streams, never crossing each other, but coruscating with a rapid interior motion, rich with chromatic displays.

Those of the second class propagate themselves from points of the compass between the north and west, toward the opposite quarters, or sometimes from the southeast, and extending themselves to the northwest: they are arch-like in form; with beaming wreaths, flashing "merry dancers," and jets of pea-green, purple, and violet light, like the spark in an exhausted receiver. But in both classes the arches are in a plane seldom deviating more than two points from the magnetic meridian. Mr. Hood has not described a vibratory motion without colors.

In the auroras seen by the American Expedition, a distinct scintillation was rare; and I observed a prismatic tinting in only a single instance. The movements of the auroral bands were so wave-like that they were at once suggestive of wind-action, although no correspondence was noted between them and the direction of the lower atmospheric currents. This effect, which I had repeated occasion to observe, heightened the resemblance of our Arctic aurora to illuminated cirro-stratus, and, I confess, always impressed me with its want of altitude.

Let me condense from my Journal and Meteorological Record a description of the aurora, as we sometimes saw it.

The 2d of February came to us with sunshine, the atmosphere in yellow light, and full of minute spiculae; our thermometers at 32°, my spirit standard at 34°, and Green's mercurial at 38°.

Drawing the finger through the mercury of our artificial horizon gave the sensation of scalding water. The evaporation and increased dryness were very perceptible: a brass clinometer, which was coated with hoar-frost, became perfectly clean on exposure to the solar ray. The haze disappeared from the southern horizon, and the sky became strikingly clear. As late as half past eight A.M., I saw the North Star in the zenith, the tail of the Bear, and stars of the third and fourth magnitude. By nine every one had gone, leaving Arcturus and Capella in possession of the field.

Between the hours of six and eight P.M., we had an interesting display of the aurora. It was of a luminous white, not much more marked than any of the isolated nebulae seen through a telescope, which it indeed resembled. This white light stretched in cumulated masses from the northwest to the southeastern horizon, forming to the northward an arch of some regularity. From the inner circumference of this great arch proceeded a series of scintillating processes, at a apparent right angles to the plane of the horizon, and constantly shifting their positions, so as to produce an effect nearly like that of the "merry dancers." To the south, however, the arch became irregular and changing; its diameter varied from five to thirty degrees, the augmentation being by a broken series of parallel bands, no one exceeding six to eight degrees.

At the period of its greatest intensity, 7h. 10 m., it enveloped Procyon and Pleiades, obscuring the larger portion of Taurus, and actually hiding Aldebaran. A process extended obliquely from about twelve degrees above the horizon to Castor and Puliux, whose brightness is sensibly dimmed. The zone then narrowed, passing about eleven degrees to the west of Polaris, and ascending in a regular arch to the northwest. It faded gradually, and by 9h. 20m. had disappeared. Neither a silk-suspended magnetic needle nor our rude electrometers detected any disturbance.

The foggy segment which forms the characteristic feature of the incipient aurora, as observed by Biot, Mairan, Lottin, and others, was in a rudimentary form visible with us. The deep bistro-colored arc, which I have arbitrarily spoken of as No. 4, is in many of its features analogous to that of the Shetland and Bossekop Observations.

The well-known aurora of Mairan begins with a dark mist or foggy cloud to the northward, not unlike the "bistro-colored segment," taking gradually an arch-like form. The visible portion of this arc soon becomes surrounded with a pale light, followed by the formation of other concentric arcs: next comes jets and colored rays from the dark part of the segment, breaking up its continuity, and indicating a general movement throughout its mass – "internal shocks," as Lardner calls them – which issue from it as flames from a conflagration.

Lottin's observations at Bossekop, in Finland, latitude 70°, which embrace no less than a hundred and forty-five exhibitions, begin with a "tinting of the constantly prevailing sea fog," the upper border of which was fringed with auroral light.

If these, and the more familiar accounts of the aurora in the middle United States, be taken as good types of this phenomenon, I would say that the matured Arctic aurora resembled their incipient stages; but that the same law of correspondence, which marks the centre of the segment in or about the magnetic axis, gave to us, situated as we were in the immediate proximity of the magnetic pole of our earth, the strange spectacle of a complete arch passing through or near the zenith, and embracing an amplitude of nearly one hundred and eighty degrees. The zone of band-like character of this auroral arch was its pervading characteristic. It seldom exceeded thirty, and was generally within ten degrees in width, a floating, waving band of nebulous illumination.

The likeness between some of the auroral appearances and a lower range of meteorological phenomena has been repeatedly noticed. The bandes polaires of Humboldt, the plaques aurorales of Lottin, the cirrocumulated resemblances of Hood and Richardson, are among these: and I have alluded more than once myself to the apparent wind-movements of our exhibitions in Lancaster Sound.

I have quoted the "fog or cloud-like segment" as forming a prominent feature in the Continental descriptions, for the purpose of introducing from my journal two anomalous exhibitions of aurora in the same connection. One was in direct conjunction with the diffracted solar ray ; the other a true daylight aurora. I gave them verbatim from my notes.

"February 7. Cold and clear : thermometer, at 8 h. 40 m. A.M., at 38°, while on the vessel's stern ; and at 42° when freely suspended by the bows outside ; my Green's spirit standard, some fifty paces from the vessel, at - 48° ; one more illustration of the local influences of ship-board, and of the irregularity of our system of registration.

"The sun was completely visible at about ten A.M. ; but his rays were subdued by a slight haziness, caused by myriads of crystallized specks that filled the atmosphere. These, when examined by my traveling Frauenhofer at two hundred diameters, gave in some cases regular hexagonal prisms, with well-defined terminations ; but this symmetry of form was generally obscured by groupings, and long oblique truncations. I have now made eight careful examinations of these crystalline spiculae at varying temperatures, when they came to us accompanied by parhelia, halos, or anomalous columns proceeding from the sun. In every case there was a decided approach to the six-sided form.

"The sun to-day exhibited an unusual phenomenon. At 10h. 20m., while very low, a column of light was observed stretching from the upper summit of its disk to an approximate height of 15°. This expanded, fan-fashioned, as it rose, and was lost by its penciled radiations blending with the illuminated sky. Thus far it did not differ materially from the vertical or crepuscular rays accompanying rudimentary forms of parhelia. But by eleven o'clock this fan-like column had enlarged to a cloudy shaft of bright yellow light, twenty to twenty-four degrees in height, and proceeding from a complete segment of illumination, which was thickly studded with cirrus clouds. The upper terminus of this column, unlike the parhelia which we had seen before, assumed a curvilinear wedge shape not unlike the section of a pear, from whose sides rose tangentially a series of penciled illuminations terminating in streaks of cloud strata.

The feature about this phenomenon of greatest interest was a distinct play of light, a series of coruscating changes resembling the scintillations of the aurora. The rays which shot out from the three-curved summit sometimes extended twelve or fifteen degrees, with a sudden movement of increased energy almost resembling ignition : then again they retired, until represented by but a few feeble points. The cloud-like segment showed in a lesser degree the same movements ; and at the periods of most active display, the vertical or fan-like shaft flashed up into more intense illumination. The diameter of this shaft at its entering base could not have been less than eighty degrees.

This singular exhibition recalled irresistibly the analogous phenomenon of the aurora, with those anomlous displays of coronae which have been referred to the diffraction of light by atmospheric vesicles or icy spiculae. I give it from my notes, as a simple detail of facts, without comment or opinion.

A daylight aurora has been described by other observers. I witnessed several, one of them interesting enough to be worth transcribing.

"About ten o'clock, going out to exercise at foot-ball, I noticed that the usual cloud-bank of the horizon had nearly cleared away at the south. One or two feathery cirri hung about the zenith, and the northern horizon retained its usual deep obscurity. This was in the course of my usual cursory examination for my weather record. Half an hour after, I observed one spot where the banking remained, attracting attention by its nearness to the sun and its well-defined segmentary character. Its margin was distinctly and regularly arched ; its tinting a peculiar purple, slightly warmed or bronzed at its margins, but deepening into a heavy brown at the line of the horizon. The centre of the segment bore south twenty degrees west (magnetic), its altitude eight degrees, nearly. Smoke and vapor from ship's fires, purple-tinted ; distant objects not very clearly visible ; atmosphere filled with ice spiculae.

"Soon from the circumference of this arch proceeded fimbriated or fringy series of purple cirri, delicately tinted at their edges, increasing with wonderful regularity, and extending in long, ray-like processes of cloud to an altitude of some twenty degrees above the horizon. Before eleven o'clock these processes had become long, stratiform illuminated clouds, beautifully marked, of a breadth, measured roughly by the eye, of four of five degrees, interrupted where they crossed the illuminated region of the sun, but every where else extending over the heavens to the south and west (true); and although still diminishing in intensity, extending nearly to the eastern quarter of the sky. By coalescing at their bases, these radiating processes augmented the size of the central segment. The intervals between them appeared, by contrast, to be artificially illuminated.

"Till now there has been no movement ; but at 11h. 20m. these cloud-like processes or radiations strikingly resembled the rays or beams of a coruscating auroral arch. Dr. Vreeland and myself witnessed repeatedly interruptions of their continuity ; then sudden shootings out, or increasings of their length ; and then a rapid and momentary formation, followed by a sudden and complete disappearance.

"At this time, too, a strange wavy movement was seen about the shorter prolongations in the neighborhood of the vertex of the mass. These resembling the rising wreaths of 'frost smoke' seen in Wellington Channel, and had an appearance almost of combustion.

"During all these phases, the cloud-like character was singularly preserved : the rays appeared to modify these processes, as light would behind our ordinary clouds. The whole exhibition was a daylight one, perfectly cloud-like, differing only in the elements of shape, movement, and radiated illumination. It was a day aurora.

"The appearance continued until twenty minutes of meridian. At 11h. 10m., when it was at its maximum, the rayed prolongations stretched nearly across the sky ; and the centre of the mass from which they emanated was fifteen degrees west from the south pole of the needle. At about the same deviation, viz., N. by E. 1-2 E., and at a rude altitude of about fifteen or twenty degrees, was an irregular cirro-cumulated cloud of the same purple tint, but not so much illuminated. From its eastern margin, rays or processes were seen stretching as high as fifty degrees, and as far as due east.

"Before the sun had reached his meridian altitude, the prolongations had become faint, and passed into detached feathery clouds, which collected at the zenith and lost the radiated arrangement altogether. The mass of cloud stratus to the south (magnetic), also, had blended with the usual bank about the horizon."

See also page 425 Vol. I., second expedition, as follows : -

"The intense beauty of the Arctic firmament can hardly be imagined. It looked close over our heads, with its stars magnified in glory and the very planets twinkling so much as to baffle the observations of our astronomer. I am afraid to speak of some of these night-scenes. I have trodden the deck and the floes, when the life of earth seemed suspended, its movements, its sounds, its colorings, its companionships ; and as i looked on the radiant hemisphere, circling above me as if rendering worship to the unseen Centre of light, I have ejaculated in humility of spirit, 'Lord, what is man that thou art mindful of him?' And then I thought of the kindly world we had left, with its revolving sunshine and shadow, and the other stars that gladden it in their changes, and the hearts that warmed to us there ; till I lost myself in memories of those who are not ; - and they bore me back to the stars again.

"The Esquimaux, like all other nomads, are careful observers of the heavenly bodies. An illustration of the confidence with which they avail themselves of this knowledge occurred while Petersen's party were at Tessieusak. I copy it from my journal of November 6.

"A number of Esquimaux sought sleeping-quarters in the hut, much to the annoyance of the earlier visitors. The night was clear ; and Petersen, anxious to hasten their departure, pointed to the horizon, saying it would soon be daylight. 'No,' said the savage ; 'when that star there gets round to that point,' indicating the quarter of the heavens, 'and is no higher than this star,' naming it, 'it will be time to harness up my dogs.' Petersen was astounded ; but he went out the next morning and verified the sidereal fact."

We now come to the proper place to show that the rounding point or that place where we might begin to descend, is at or near the latitude of 70°. To show this we shall bring a variety of arguments forward, either one of which would seem to be conclusive.

The first freezing of New Hampshire commences on the top of Mount Washington, from which we conclude that this is the coldest place in the state. Now Parry, Ross, Franklin, and Kane all testify to the fact that the first freezing in the Arctic regions commences south of 70°, showing this to be the coldest part ; and that north as well as south of this point it is warmer. Dr. Kane when at Rensselaer Harbor found it warmer than Godfrey did 250 miles south of him.

The tides are found to commence at 70°, when their revolution is complete. They sometimes, though seldom, go as far as 80°, at which latitude Morton met with tides coming south.

Dr. Kane, or Morton, discovered three kinds of whales, walruses, seal, bears, and deer, around the open ocean, and the wind blew three days from the north, and it brought no ice, and Parry traveled up on the ice the east side of Greenland to 82° North, and found open pools and rotten ice, and met birds coming all from the north, and he rested the last day, and it was the warmest day he had while on his journey. See Kane, Vol. I., p. 288 and onward on tides, etc.

"At the pitch of the cape the ice-ledge was hardly three feet wide ; and they were obliged to unloose the dogs and drive them forward alone. Hans and he then tilted the sledge up, and succeeded n carrying it past the narrowest place. The ice-foot was firm under their tread, though it crumbled on the verge.

"The tide was running very fast. The pieces of heaviest draught floated by nearly as fast as the ordinary walk of a man, and the surface-pieces past them much faster, at least four knots. On their examination the night before, the tide was from the north, running southward, carrying very little ice. The ice which was now moving so fast to northward, seemed to be the broken land-ice around the cape, and the loose edge of the south ice. The thermometer, in the water, gave 36°, seven degrees above the freezing-point of sea water at Rensselaer Harbor.

"They now yoked in the dogs, and set forward over the worst sort of mashed ice, for three quarters of a mile. After passing the cape, they looked ahead and saw nothing but open water. The land to the westward seemed to overlap the land on which they stood, a long distance ahead ; all the space between was open water. After turning the cape, - that which is marked on the chart as Cape Andrew Jackson, - they found a good smooth ice-foot in the entering curve of a bay, since named after the great financier of the American Revolution, Robert Morris. It was glassy ice, and the dogs ran on it full speed. Here the sledge made at least six miles an hour. It was the best day's travel they made on the journey.

"After passing four bluffs on the bottom and sides of the bay, the land grew lower ; and presently a long low country opened on the land-ice, a wide plain between large headlands, with rolling hills through it. A flock of Brent geese were coming down the valley of this low land, and ducks were seen in crowds upon the open water. When they saw the geese first, they were apparently coming from the eastward ; they made a curve out to seaward, and then, turning, flew far ahead over the plain, until they were lost to view, showing that their destination was inland. The general line of flight of the flock was the northeast. Eiders and dovekies were also seen, and were very numerous, hundreds of them squealing and screeching in flocks. they were so tame that they came within a few yards of the party. Flying over head, their notes echoing from the rocks, were large white birds, which they took for burgomasters. Ivory gulls and mollemokes were seen farther on. They did not lose sight of the birds after this, as far as they went. The ivory gulls flew very high, but the mollemokes alit, and fed on the water, flying over it well out to sea, as we had seen them do in Baffin's Bay. Separate from these flew a dingy bird unknown to Morton. Never had they seen the birds so numerous ; the water was actually black with dovekies, and the rocks crowded.

"The part of the channel they were coasting was narrower, but as they proceeded it seemed to widen again. There was some ice, arrested by a bend of the channel on the eastern shore ; and, on reaching a low gravel point, they saw a projection of land shut them in just ahead to the north. Upon this ice numerous seal were basking, both the netsik and ussuk.

"To the left of this, toward the West Land, the great channel (Kennedy Channel) of open water continued. There was broken ice floating in it, but with passages fifteen miles in width and perfectly clear. The end of the point - 'Gravel Point,' as Morton called it, - was covered with hummocks and broken ice for about two miles from water. This ice was worn and full of gravel. Six miles inland, the point was flanked by mountains.

"A little higher up, they noticed that the pieces of ice in the middle of the channel were moving up, while the lumps near the shore were floating down. The channel was completely broken up, and there would have been no difficulty in a frigate standing anywhere. The little brig, or 'a fleet of her like,' could have beat easily to the northward. The wind blew strong from the north, and continued to do so for three days, sometimes blowing a gale, and very damp, the tops of the hills becoming fixed with dark foggy clouds. The damp falling mist preventing their seeing any distance. Yet they saw no ice borne down from the northward during all this time ; and, what was more curious, they found on their return south, that no ice had been sent down during the gales. On the contrary, they found the channel perfectly clear from shore to shore.

"*Thursday, June 22.* - They camped at 8:30 A.M., on a ledge of low rock, having made in the day's journey forty-eight miles in a straight line. Morton thought they were at least forty miles up the channel. The ice was here moving to the southward with the tide. The channel runs northwardly, and is about thirty five miles wide. The opposite coats appears straight, but still sloping, its head being a little to the west of north. The shore is high, with lofty mountains of sugar-loaf shape at the tops, which, set together in ranges, looked like piles of stacked cannon balls. It was too cloudy for observations when they camped, but they obtained several higher up. The eider were in such numbers here that Hans fired into the flocks, and killed two birds with one shot.

"*Friday, June 23.* - In consequence of the gale of wind, they did not start till 12:30 midnight. They made about eight miles and were arrested by the broken ice of the shore. Their utmost efforts could not pass the sledge over this ; so they tied the dogs to it, and went ahead to see how things looked. They found the land-ice growing worse and worse, until at last it ceased, and the water broke directly against the steep cliffs.

"They continued their course overland until they came to the entrance of the bay, whence they could see a cape and an island to the northward. They then turned back, seeing numbers of birds on their way, and leaving the dogs to await their return, prepared to proceed on foot. This spot was the greenest that they had seen since leaving the headlands of the channel. Snow patched the valleys, and water was trickling from the rocks. Early as it was, Hans was able to recognize some of the flower life. He eat of the young shoots of the lychnis, and brought home to me the dried pod (*siliqua*) of a hesperis, which had survived the water and tear of winter. Morton was struck with abundance of little stone-crops, 'about the size of a pear.' I give in the appendix his scanty list of recognized but not collected plants.

"*Friday and Saturday, June 23 and 24.* - At 3 A.M. they started again, carrying eight pounds of pemmican and two of bread, besides the artificial horizon, sextant, and compass, a rifle, and the boat-hook. After two hours walking the travel improved, and on nearing a plain about nine miles from where they had left the sledge, they were rejoiced to see a she-bear and her cub. They had tied the dogs securely, as they thought, but Toodla and four others had broken loose and followed them, making their appearance within an hour. They were thus able to attack the bear at once.

"Hans, who to the simplicity of an Esquimaux united the shrewd observation of a hunter, describes the contest which followed so graphically that I try to engraft some of the quaintness of his description upon Mr. Morton's report. The bear fled ; but the little one being unable to keep ahead of the dogs or to keep pace with her, she turned back, and, putting her head under its haunches, threw it some distance ahead. The cub safe for the moment, she would wheel round and face the dogs, so as to give it a chance to run away ; but it always stopped just as it alighted, till she came up and threw it ahead again ; it seemed to expect her aid, and would not go on without it. Sometimes the mother would run a few yards ahead, as if to coax the young one up to her, and when the dogs came up she would turn on them and drive them back ;

then, as they dodged her blows, she would rejoin the cub and push it on, sometimes putting her head under it, sometimes catching it in her mouth by the nape of the neck. For a time she managed her retreat with great celerity, leaving the two men far in the rear. They had engaged her on the land-ice ; but she led the dogs in-shore, up a small stony valley which opened into the interior. But after she had gone a mile and a half her pace slackened, and, the little one being jaded, she soon came to a halt. The men were then only half a mile behind ; and, running at full speed, they come up the where the dogs were holding her at bay. The mother never went more than two yards ahead, constantly looking at the cub. 'Never,' said Morton, 'was an animal more distressed.' She would stretch her neck and snap at the nearest dog with her shining teeth, whirling her paws like the arms of a windmill. If she missed her aim, not daring to pursue one dog lest the others should harm the cub, she would give a great roar of baffled rage and go on pawing, and snapping, and facing the ring, grinning at them with her mouth stretched wide.

"When the men came up, the little one was perhaps rested, for it was able to turn around with her dam, no matter how quick she moved, so as to keep always in front of her belly. The five dogs were all the time frisking and her actively, tormenting her like so many gad-flies ; indeed, they made it difficult to draw a bead on at her without killing them. But Hans, lying on his elbow, took a quiet aim and shot her through the head. She dropped and rolled over dead without moving a muscle.

"The dogs spring toward her at once ; but the cub jumped upon her body and reared up, for the first time growling hoarsely. They seemed quite afraid of the little creature, she fought so actively and made so much noise ; and, while tearing mouthfuls of hair from the dead mother, they would spring aside the minute the cub turned toward them. The men drove the dogs off for a time, but were obliged to shoot the cub at last, as she would not quit the body.

"Hans fired into her head. It did not reach the brain, though it knocked her down ; but she was still able to climb on her mother's body and try to defend it still, 'her mouth bleeding like a gutter-spout.' They were obliged to dispatch her with stones.

"After skinning the old one, they gashed its body, and the dogs fed upon it ravenously. The little one they cached for themselves on the return ; and with difficulty taking the dogs off, pushing on, crossing a small bay which extended from the level ground and had still some broken ice upon it. Hans was tired out, and was sent on shore to follow the curve of the bay, where the road was easier.

"The ice over the shallow bay which Morton crossed was hummocked, with rents through it, making very hard travel. He walked on over this, and saw an opening not quite eight miles across, separating the two islands, which I have named after Sir John Franklin and his comrade Captain Crozier. He had seen them before from the entrance of the larger bay, - Lafayette Bay, - but had taken them for a single island, then channel between them not being then in sight. As he neared the northern land, at the east shore which led to the cape (Cape Constitution), which terminated his labors, he found only a very small ice-foot, under the lee of the headland and crushed up against the side of the rock. He went on, but the strip of land-ice broke more and more, until about a mile from the cape it terminated altogether, the waves breaking with a cross sea directly against the cape. The wind had moderated, but was still from the north, and the current ran up very fast, four or five knots perhaps.

"The cliffs were very high ; at a short distance they seemed about two thousand feet ; but the crags were so overhanging that Morton could not see the tops as he drew closer. The echoes were confusing, and the clamor of half a dozen ivory gulls, who were frightened from their sheltered nooks, was multiplied a hundred-fold. The mollemokes were still numerous, but he now saw no ducks.

"He tried to pass round the cape. It was in vain ; there was no ice-foot, and, trying his best to ascend the cliffs, he could get up but a few hundred feet. Here he fastened to his walking-pole the Grinnell flag of the *Antarctic* - a well-cherished little relic, which now had followed me on two Polar voyages. This flag had been saved from the wreck of the United States sloop-of-war *Peacock*, when she was stranded off the Columbia River ; it had accompanied Commodore Wilkes in his far-southern discovery of an Antarctic continent. It was now its strange destiny to float over the highest northern land, not only of America, but of our globe. Side by side with this were our Masonic emblems of the compass and the square. He let them fly for an hour and a half from the black cliff over the dark rock-shadowed waters, which rolled up and broke in white caps at its base.

"He was bitterly disappointed that he could not get round the cape, to see whether there was any land beyond ; but it was impossible. Rejoining Hans, they supped off their break and pemmican, and, after a good nap, started on their return on Sunday, the 25th, at 1:30 p.m. From Thursday night, the 22nd, up to Sunday at noon, the wind had been blowing steadily from the north, and for thirty-six hours of the time it blew a gale. But as he returned, he remarked that the more southern ice toward Kennedy Channel was less than it had been when he passed up. At the mouth of the channel it was more broken than when he saw it before, but the passage above was clear. About half way between the farthest point which he reached and the channel, the few small lumps of ice which he observes floating - they were not more than half a dozen - were standing with the wind to the southward, while the shore-current or tide was driving north.

"His journal of Monday, 26th, says, 'As far as I could see, the open passages were fifteen miles or more wide, with sometimes mashed ice separating them. But it is all small ice, and I think it either drives out to the open space to the north, or rots and sinks, as I could see none ahead to the far north.'

"The coast after passing the cape, he thought, must trend to the eastward, as he could at no time when below it see any land beyond. But the west coast still opened to the north ; he traced it for about fifty miles. The day was very clear, and he was able to follow the range of mountains which crown it much farther. They were very high, rounded at their summits, not peaked like those immediately abreast of him ; though, as he remarked, this apparent change in their character might be referred to distance, for their undulations lost themselves like a wedge in the northern horizon.

"His highest station of outlook at the point where his progress was arrested he he supposed to be about three hundred feet above the sea. From this point, some six degrees to the west of north, he remarked in the farthest distance a peak truncated at its top like the cliffs of Magdalena Bay. It was bare at its summit, but striated vertically with protruding ridges. Our united estimate assigned to it an elevation of from twenty-five hundred to three thousand feet. This peak, the most remote northern land known upon our globe, takes its name from the great pioneer of Arctic travel, Sir Edward Perry.

"The range with which it was connected was much higher, Mr. Morton thought, than any we had seen on the southern or Greenland side of the bay. The summits were generally rounded, resembling, to use his own expression, a succession of sugar-loaves and stacked cannon-balls declining slowly in the perspective. I have named these mountains after the name of the lady sovereign under whose orders Sir John Franklin sailed, and the prince her consort. They are familiar in their features to those of Spitzbergen ; and, though I am aware how easy it is to be deceived in our judgment of distant heights, I am satisfied from the estimate of Mr. Morton, as well as from our measurements of the same range farther to the south, that they equal them in elevation, 2,500 feet.

"Two large indentations broke in upon the uniform margin of the coast. Everywhere else the spinal ridge seemed unbroken. Mr. Morton saw no ice."

The following extract is taken from Barrows, and English explorer who accompanied Franklin in several expeditions to the north : -

"Parry, in his voyage to Spitzbergen, went on the ice to 83° north, in longitude 19° 25' east. He came to the broken ice and open pools of water. Ice became so rotten he could travel no farther. Here he found the birds coming up from the north, and the wind blowing from the north a breeze that made the day he rested the warmest while on his journey."

It is now a generally received opinion that there is an open Polar Sea at the north pole. The evidence on this point received since the first Grinnall expedition has been so abundant as to leave no doubt of its existence. It has been a topic of theory for over two hundred years. Water was seen to the eastward of the most northern cape of Novaia Zemlia as far back as 1596. In this year Barentz, an authority certainly reliable, speaks of the increasing warmth as he left land to the north of 77° ; and the sea north of Spitzbergen is said by the early Dutch to have been as warm as the sea of Amsterdam.

Kane, in his first book, in speaking of the combination of warmth and cold in Baffin's Bay says (page 149), *"All this is attempered by the warm glazing of a tinted atmosphere. The sky of Baffin's Bay, though but eight hundred miles from the Polar limit of all northernness, is a s warm as the bay of Naples after a June rain. What artist, then, could give this mysterious union of warm atmosphere and cold landscape?"*

That the rounding point is near 70° is shown again by the compass. Kane tells us on page 282 of his first volume on his second expedition, that Morton's party, when beset with icebergs dangerous to pass, would sometimes attempt to find new routes. *"This,"* he says, *"was a tedious and dangerous alternative, as the compass, their only guide, confused them by its variation."*

We have also the evidence of Barrows that when at latitude 77°, the compass became useless ; and at a point still farther north, the needle turned directly round toward the south. Parry, and also Sontag, both testify to the same unaccountable fact.

Now it will not be admitted that they reached the pole and were descending the opposite side! We must yet wait for an explanation of so unnatural a result, if our theory be not correct.

But on our reasoning, it is at once simple and natural ; the earth being a hollow sphere, with air within pressing to the centre, in every position ; the air itself supported by the element of space, having its connection with inside air at the northern and southern extremities ; the inside earth being formed of oceans and continents, the same as the outside ; there will then be the same northern and southern magnetic attraction on the inside as on the outside. Here we have two equal opposing influences approaching each other, and it is evident, at their place of meeting, they will neutralize each other.

We will illustrate this by taking three large balls, - powerful magnets, - one weighing one hundred pounds, the others fifty each. We place the largest one in a northern position, the two others in the northeast and northwest respectively. If we now locate a compass in the southern position, the needle points north. But if we remove the compass to the large ball, it will vary so as to destroy its utility. so at 70°, or thereabouts, the compass becomes useless, as witnessed by Perry, Ross, Franklin, Kane, and every other northern traveler of note.

The traveler to the Franconia Notch, in New Hampshire, never fails to notice the many novel phenomena produced there by reflection, and other natural means. A description of a few of these effects may not be out of place here, especially as they will vividly illustrate the principle we wish to show. At the foot of the "Old Man of the Mountain," is the "Basin," a small pond nearly circular in form, and about half a mile in diameter.

When the sun is nearing the meridian, this mountain, with the profile perfect, casts its full shadow directly into the bosom of the lake. If we now take a mirror, holding it in a certain position, this shadow is reflected into the glass distinctly. If we now take several convex lenses of different colors, placing one upon the other, and look through them into the mirror, still holding it in the same position, we see the same in so many different colors. But, principally, if we look at the sun, or its reflection, we see him in the different colors, or so many different suns.

At times the sun's rays are reflected upon the mountain, which play a thousand antics on the Old Man's sides, with every wave of the lake, producing the aurora in a perfect though miniature form. The same effect is seen on the stone walls of the "Pool," from eleven to three o'clock, as every traveler there has observed after the sun has left the eastern side.

The principle on which this would be explained, will apply to the Aurora Borealis ; the icebergs around the circle near the extremities, borne by the wakes of the ocean, forming so many reflecting glasses to throw off the sun's rays in every direction. These rays are again reflected by the higher clouds, or the cirri, and brought within view of those dwelling toward the extremity of from which they proceed. It will be noticed that these icebergs are in every conceivable shape and form, and the colors are produced by their prismatic construction.

It is probable that when Kane saw the three suns "coming to greet him," he stood in the foci of rays of light passing through convex icebergs, so that the suns would be, apparently, visible to him.

It is generally believed that within, this world is an enormous burning mass ; that we live on a crust beneath which eternal liquid fires are constantly burning, sometimes melting the earth and the everlasting rocks, and at other times cooling off, the escaping gases causing earthquakes, volcanoes, hot springs, and a variety of causes, and call the absurdity proved beyond controversy.

It seems that such a belief would appear more ludicrous than even our theory of its perforation. And though it is superfluous to combat such a belief, we shall do it, and meanwhile discover our own philosophy of volcanoes and earthquakes.

That there is fire within the earth we do not deny. The sublime and often disastrous eruptions of Vesuvius are too well known to need description. But the fact that Vesuvius and Etna never burn rapidly together is not, perhaps, so well known. It will be well, however, to bear in mind that such is the fact ; for when one is in full blast, the other appears to cool off, or to burn slightly.

This singular action has never, to my knowledge, been explained, hence I venture my opinion in regard to it, as it is illustrative of several important points. Etna is situated in the island of Sicily, two hundred miles from Vesuvius. It is one of the most celebrated volcanoes in the world, and the longest known in history. It is about eighty miles in circumference, and nearly eleven thousand feet in height. Vesuvius is situated six miles from the city of Naples, and is three thousand nine hundred feet high.

Taking these two volcanoes as examples by which to prove the laws that govern others, we find that Vesuvius has frequently overflown its banks, at which times the amount of pressure sustained in its magnificent hollow cone, is not less than thirty million tons, allowing a weight of sixty-four pounds to the square foot. This tremendous pressure is produced by the gases beneath. The fires within, destroy the oxygen of the air, forming gases which rise to seek a vent. Here the accumulated lava resists its attempt ; two antagonistic forces meet, and the lava, being the weaker, rises in the vast funnel, above the gas, till it reaches the tapis, where a portion is discharged, and also vast quantities of the gas, till it can longer sustain the lava, which, in turn, forces the gas back into the earth. It now fills every pore, permeating to the surface even, sometimes escaping from deep fissures, but oftener taken up by the multitudinous roots of the trees and plants, thereby going to the support of vegetable life. Here, this great chemical operation is continued, the plant appropriating to its own use a requisite amount, while any superfluity seeks the air through every leaf and twig. It now escapes to the clouds, and condensed with other gaseous particles, to fall again in showers. In this connection, it may be well to remark, that the outside rind or bark of the trees is similar to the skin of animals, admitting the perspiration to escape in a similar manner. In performing this office, it throws off the oxygen so necessary to the support of animal life.

These gases, escaping from volcanoes beneath the earth's surface, often come in contact with water, either n the inside or on the outside. The gases, being a lighter element, seek to the top. No one has failed to notice the small bubbles often seen rising to the top of ponds and lakes, which are caused by the pressing up of the subterraneous gases.

This gas still rises through the air, with the water, as fast as it evaporates, according to the magnitude of the rays of light thrown upon the earth's surface in every position, which create heat ; rarefying the air, and causing the gases to rise till they are condensed by the coolness of the upper regions, and fall again in the shape of rain.

To return to the explanation of the cause of Etna's remaining quiet while Vesuvius is belching forth its fury ; we say, when the latter begins to burn, it draws all the gases in its vicinity to itself, until its combustible particles are mostly consumed ; when a reaction ensues, and Etna, in turn, whose fires have lain dormant for a time, revives, drawing the gases from Vesuvius and appropriating them to its own use. To make this more clear, we will suppose we have two stoves in a perfectly air-tight room, with different connections to the outside air. If we now light a fire in one, it will be fed with the air in the room, and will soon draw it all to itself. If we now undertake to light a fire in the other stove, it will be but a poor attempt, as the air will continue to feed the former stove as long as it contains anything to support combustion ; when this is exhausted, and it begins to cool off, a fire may then be lighted in the other stove, and in a few minutes it will so gain the ascendency over the first as to render it difficult to kindle one in that. These two volcanoes being connected in a similar manner by subterraneous gases, operate on the same principle.

The cause of the various hot springs, is owing to the hot gases from volcanic fires, coming into contact with their waters. And the hot gases issuing from the earth, in the vicinity of Bear River, undoubtedly come from a volcano in the inside, there being none on the outside within a great distance. The oozing up of these hot gases is more frequent in volcanic regions than in other places, which shows that they proceed from volcanic influence. Yet they are not infrequently found in countries free from volcanoes, as in Utah. Here there are several localities where such occurrences are found, some of which are exceedingly poisonous, such as the one known as "Dograta," so highly charged with poisonous matter, that an animal thrown into it dies in a short time. These gases must all proceed from some internal volcano.

Earthquakes are caused by the eruption of volcanoes, and the lava of volcanoes passing down to the under ocean from the upper continent or island, and from the same passing from the under continents or islands to the outer ocean. This disastrous consequence happens when a volcano has burned down, on either side, to the waters of the ocean on the opposite side ; when this occurs, more gases and steam are produced ; fire and water cannot dwell together, and a powerful concussion, shaking the whole earth in the vicinity, sometimes causing it to open and swallow whole cities, is the result.

In the South Pacific Ocean there is a stream drifting northeast known as the Antarctic Drift Current. Its course is nearly contrary to the equatorial current which flows west ; no satisfactory explanation has ever been given to account for it, and philosophers and geographers say, "cause unknown." There is also on the western side of the Atlantic, commencing in the Gulf of Mexico, and running northeast to Baffin's Bay, another current known as the

Gulf Stream, a stream so powerful as to baffle the tide in the Gulf of Mexico and Caribbean Sea. Several weak attempts have been made at its explanation, but none of sufficient weight to be worthy of mention. But the greatest current and the one most generally understood, is the equatorial current. This commences in the Eastern Pacific Ocean, and flows both sides of the equator through the Pacific, without interruption, as there is nothing to break its force. But when it reaches the western side of Asia and Africa, and during its passage through the Indian and Atlantic Oceans, it is broken and variable, owing to the large continents which continually obstruct its passage. It is universally admitted that the cause of this current is the diurnal motion of the earth, super-induced by the friction between the continent and the water as the earth rotates east. Now supposing before, the ocean to be four miles deep, then the two first miles in depth will form an upper current setting west, and the two lower miles are an under current setting east, and the two lower miles having the two upper miles pressing upon it, will press with as much power as though both currents were going the same way. At the Isthmus of Darien, the Atlantic is but about forty miles from the Pacific, and that they may here be connected by a subterranean passage is not improbable. Now, the Pacific being fifteen feet higher than the Atlantic, and the tide in the former rising twenty-four feet higher than in the latter, this gives the Pacific a pressure of forty feet over the Atlantic. This would be sufficient to force the water from the under current of the Pacific through this connection with great power, sufficient at least to cause the Gulf Stream. The water in the lower currents is saltier and warmer than at the top, which will account for its superior warmth and saltiness.

In Baffin's Bay there are two currents setting north and four setting south, which meet at what is known to Arctic travelers as the "Centre Pass," a portion of the bay covered with floating ice.

There is at this place, in all probability, a subterranean passage to an under ocean absorbing the water from these different currents, as it approaches. A similar passage from the inside ocean at the southern extremity may be the cause of the Antarctic current before alluded to, in the southern Pacific.

All the seas and lakes that lie below the level of the ocean, such as the Caspian Sea, which is eighty-four feet below, and having the rivers Volga, Kour, Terek, and Ural, carrying into it 900,000,000 square feet of fresh water every hour ; also Lake Tiberias lying six hundred feet below the ocean, with the large rivers, Sihoon and Amoo, emptying into it, as well as the Dead Sea lying 1,316 feet below, with the River Jordan emptying into it, all these, no one having a visible outlet, are connected with the under or inside ocean. The fresh water of these mighty rivers pouring into the lakes press heavily upon the waters of the under ocean and find their escape into it. This creates a current between the seas and the under ocean ; the salt waters of the latter pressing up are the cause of these lakes being so salt.

In conclusion I will only say that I believe all the planets are constructed similar to ours ; and that Saturn, instead of being possessed of rings, as is usually believed, is placed in a position with one extremity toward us into which we look through the other, giving all the appearance of a ring. The two are produced by the reflection of light. I shall have occasion to speak farther in regard to this planet hereafter, and simply allude to it in this place as particularly illustrative of views before advanced.

Though the principles now set forth may seem improbable to many, I would call attention to them, and present these new theories as subjects for their candid and impartial investigation, believing they will be found more strictly in accordance with the laws of nature than the old dogmas of philosophy so generally received. I have thought best to give, in an appendix, one of Dr. Kane's Lectures entire, which will be found rich in information and abounding in arguments in favor of our position.

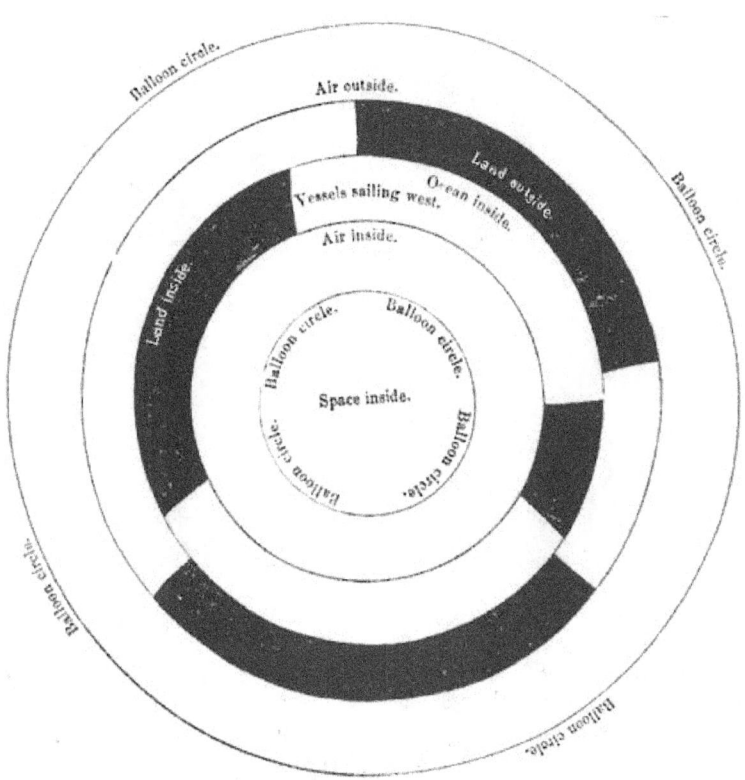

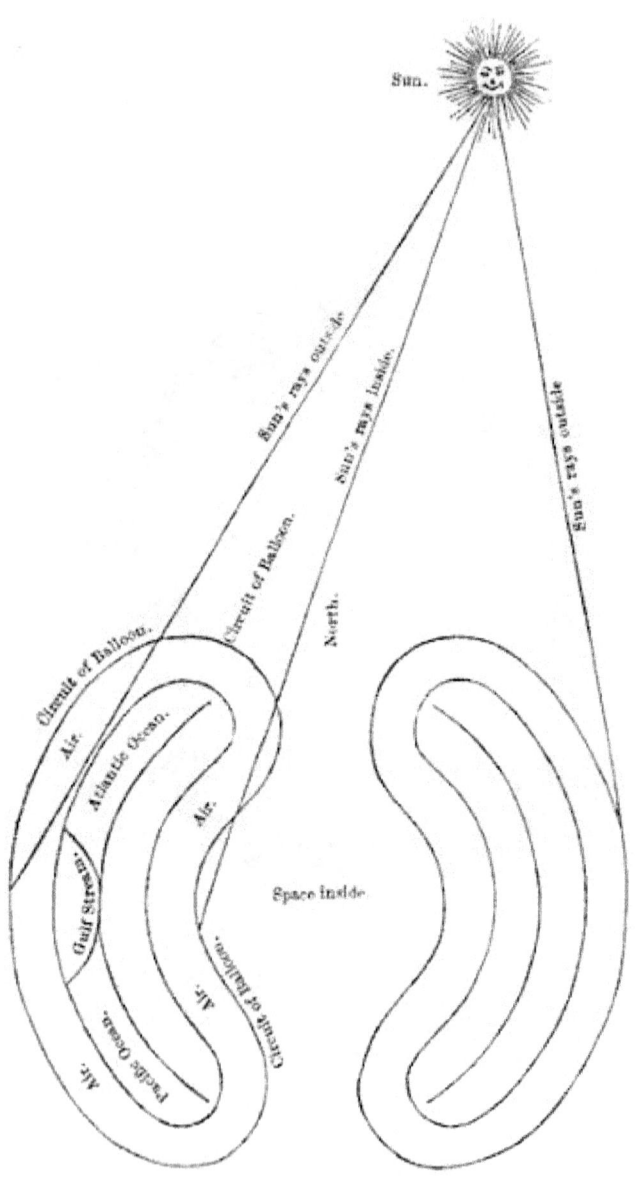

COSMOGONY: OR THOUGHTS ON PHILOSOPHY

DR. KANE'S LECTURE

Lecture on the Access to an Open Polar Sea in connection with the Search after Sir John Franklin and his Companions, read before the American Geographical and Statistical Society at its regular monthly meeting, by Dr. Kane, December 14, 1852:

The north pole, the remote northern extremity of our earth's axis of rotation, is regarded, even by geographers, with that mysterious awe which envelopes the inaccessible and unknown.

It is shut out from us by an investing zone of ice ; and this barrier is so permanent, that successive explorers have traced its outline, like that of an ordinary sea-coast.

The early settlements of Iceland, and their extensions to Greenland, as far back as 900 AD, indicated a protruding tongue of ice from the unknown north, along the coast of Greenland. I must express a doubt if the early voyages of Cabot, and Frobisher, and the Cortereals did more than establish detached points of this line. The voyages, however, of the Basque and Biscayan fisherman, about 1575, to Cape Breton, made us aware of a similar ice-raft along the coasts of Labrador to the north ; and the commercial routes of the old Muscovy Company, aided by the Dutch and English whalers, extended this across the Spitzbergen, and thence to the regions north of Archangel, in the Arctic Seas. The English navigators of the days of Elizabeth, the "notable worthys of the Northe Weste Passage," spoke of a similar ice-raft up Baffin's and Hudson's bays, and the Russo-Siberians gave us vaguely a girding-line of ice, which protruded irregularly from the Asiatic and European coasts into the Polar Ocean. Lastly,

Cook proved that the same barrier continued across Behring's Straits as high as 70° 44' north.

From all this it appeared that the approaches to the pole were barricaded with solid ice. We owe the march of modern discovery, especially stimulated by the search after its great pioneer, Sir John Franklin, our ability accurately to define nearly all the coasts of a polar sea, if not to lay down the no less interesting coast of a grand continuous ice-border that encircles it. It is worthy of remark, that this ice, although influenced by winds, currents, and deflecting land masses, retains through the corresponding period of each successive year a strikingly uniform line.

During the winter and spring, from October to May, or eight months of the year, it may be found travelling down the coast of Labrador almost to Newfoundland, blockading the approaches into Hudson's Bay, and cementing into one great mass the numberless outlets which extend from it and Baffin's Bay to the unknown coasts of the north.

Influenced by the earth's rotation, this ice accumulates toward the westward, leaving an uncertain passage along the eastern waters of Baffin's Bay, after which it resumes its march along the eastern coast of Greenland, shutting in that extensive region appropriated to the interesting legend, or that meteorological myth, as it has been designated by Humboldt, of "Lost Greenland." Its next course is to the northeast, sometimes enveloping Iceland ; and thence, extending to the east by Jan Meyen's Land and Spitzbergen, it crosses the meridian of Greenwich at some point between the latitudes of 70° and 73°.

I now call your attention to a remarkable feature in this great ice coast line. Upon reaching a longitude of about 70° east, it suddenly turns toward the north, forming a marked indentation as high as latitude 80° ; then, coming again to the southeast until it reached Cherie Island, it continues with a varying line to the unexplored regions north of Nova Zembla.

This indentation or sinuosity, best known as the old "Fishing Bight" of the Greenland Seas, is undoubtedly due to the thermal influences of the Gulf Stream. We know that the coasts of Nova Zembla feel the influences of its waters ; and Petermann, and many others, guided by the projected curves of Dove, suppose that its heated current is deflected by that peninsula, so as to impress the polar ice to a greater degree of northing than on any other part of our globe. It would be important to the objects of my communications, that I should trace this ice throughout its entire extent ; but I have not the means of doing so with exactness. Barentz, in 1596, was arrested by ice in latitude 77° 25', upon the meridian of 70° east. Prontschicheff met the same rebuff at the same height 30° further west. Anjou, Matieuschinm and Wrangell found it in a varying belt along the Asiatic coast, but fifty miles in width.

The enterprise of our American whalers has also traced this ice across Behring's Straits, as high as latitude 72° 40' ; and it is probable that Herald Island, in latitude 71° 17', is a part of a great island chain, continued from Cape Yacan to Banks' Land and the Parry Island ; an archipelago whose northern faces are yet unexplored, but which undoubtedly serve as a cluster points of ice cementation, and abounds more or less with polar ice at all seasons of the year.

We have now followed, throughout its entire circuit, this immense investing body. The circumpolar ice, as I will venture to name it, may be said to bound an imperfect circle of 6000 miles in circumference with a rude diameter of 2000 miles, and an area, if we admit its continuity to the pole, one third larger than the continent of Europe. But theory has determined that this great surface is not continuous. It is annulus, a ring surrounding an area of open water, - the Polynya, or Iceless Sea.

Polynya is a Russian word, signifying an open space ; and is used by the Siberians to indicate the occasional vacancies which occur in a frozen water surface. Although such a vacancy, as applied to a polar sea, is generally recognized to exist, it is right for me to state that this opinion is not based upon the results of exploration. It is due rather to the well-elaborated inductions of Sabine and Berghaus, and especially of our accomplished American hydrographer, Lieutenant Maury.

The North Polar Ocean is a great mediterranean, draining the northern slopes of three continents, and receiving the waters of an area of 3,751,270 miles. Indeed, the river systems of the Arctic Sea exceed those of the Atlantic.

The influences of the congelation too, aided by the diminished intensity and the withdrawal of the solar ray, increase the atmospheric precipitation, and probably diminish the compensating evaporation. Yet this position calls for further investigation to establish it absolutely ; for recent experiments show that even in the dark hours of winter, and at temperatures of fifty degrees below zero, evaporation goes on at a rapid rate. That it holds, however, in general terms,. is evident from the inferior specific gravity of the arctic waters. They are less salt than those of more equatorial regions. Their average specific gravity (1.0265) indicates about 3.60 percent of saline matter.

The atmospheric precipitation extending to the adjacent land-slopes, the melting of the snows and accumulated glacial material, and the flood of the great Siberian rivers, are sufficient to account for this.

With such sources of supply, it is evident that this surcharged basin must have an outlet, and its contents a movement independent of the laws of currents generally operative, which would determine them toward the equator.

The avenues of entrance to and egress from the polar basin are but three : Behring's Straits, and estuaries of Hudson's and Baffin's bays, and the interval between Greenland and Norway, upon the Atlantic Ocean, known as the Greenland Sea. In Behring's Straits, it is probable, from imperfect observations, that the current sets during a portion of the year from the Pacific to the Arctic Sea, with a velocity varying from one to two and a half knots per hour. Neither the soundings nor the diameter of this strait indicate any very large deep-sea discharge in the other direction. The Gulf Stream, after dividing the Labrador current, has been traced by Professor Dove to the upper regions of Nova Zembla ; so that Baffin's Bay and the Hudson and Greenland seas, constitute the only uniform outlet to the polar basin.

It is by these avenues, then, that the enormous masses of floating ice, with the deeply-immersed bergs, and still deeper belt of colder water, are conveyed outward. Underlying the Gulf Stream, whose waters it is estimated at least to equal in volume, the vast submerged icy river flows southward to the regions of the Caribbean. The recent labors of the United States Coast Survey and National Observatory have developed and confirmed and previously-broached idea of a compensating system of polar and tropical currents ; and we are prepared to consider these colder streams as equalizers to the heated areas of the tropical latitudes, and analogous in cause and effect to the recognized course of the atmospheric currents.

In fact, Dove, Berghaus, and Petermann, three authorities entitled to the highest respect, recognize for the Arctic Ocean a system of revolving currents, whose direction during summer is from north to south, and during winter the reverse, or from the south to the north. The isotherms of Lieutenant Maury (projected by Professor Flye) point clearly to the same interesting result. Contrasting these great movements of discharge and supply with the surface actions, we find during the summer months a movement along the northern coasts of Russia, clearly from east to west, from Nova Zembla westwardly and southwardly to Spitzbergen, where, after an obscure bifurcation, it is met by a great drift from the north, and carried along the coast of Greenland, in a large body known as the East Greenland current. The observations collected by Lieutenant Commanding De Haven, show that this stream is deflected around Cape Farewell, passing up the Greenland coast to latitude 74° 76' ; where, after coming to the western side of the bay, it passes along the eastern coast of America, even to the Capes of Florida. During the winter, when the great rivers of Siberia and America lose their volume by the action of the frost, a current has been noted from the Faroe Islands, north and east along the Asiatic coasts toward Behring's Straits. And then it is that the great surface ice, formed upon the coasts of Asia, gives place to a warmer stream, and the heated waters of the Gulf current bathe and temper the line of the Siberian coast.

All these go to prove that the polar basin is not only the seat of an active supply and discharge, but of an intestine circulation independent of either ; while the intercommunication of the whales between the Atlantic and Pacific, as shown by Maury, proves that the two oceans are united.

Admitting the important fact of a moving, open sea, the recognized equalization of temperatures attending upon large water masses follows of course. But is the Arctic Sea, in fact, an unvaried expanse of water? For if it be not, the excessive radiation and other disturbing influences of land upon general temperature are well known. It is, I think, and open sea. And an argument may be deduced for this belief from the icebergs.

The iceberg is an offcast from the polar glacier, and needs land as an essential element in its production, - as much so as a ship the dockyard on which she is built. From the excessive submergence of these great detached masses, they may be taken as reliable indices of deep-sea currents, while their size is such that they often reach the latitudes of the temperate zone before their dissolution. Now it is a remarkable fact that these huge ice-hulks are confined to the Greenland, Spitzbergen, and Baffin seas. Throughout the entire circuit of the Polar Ocean, almost seven thousand miles of circumscribing coast, we have but forty degrees which is ever seen to abound in them.

A second argument, bearing upon this, is found in the fact that a large area of open water exists, between the months of June and October, in the upper parts of Baffin's Bay. This mediterranean Polynya is called by the whalers the North Water. After working through the clogging ice of the intermediate drift, you pass suddenly into an open sea, washing the most northern known shores of our continent, and covering an area of 90,000 square miles.

The iceless interval is evidently caused by the drift having travelled to the south without being reinforced by fresh supplies of ice ; and the latest explorations from the upper waters of this bay speak of avenues thirty-six miles wide extending to the north and east, and free. The temperature os this water is sometimes 12° above the freezing point ; and the open bays or sinuosities, which often indent the Spitzbergen ice as high as 81° north latitude, have been observed to give a sea-water temperature as high as 40°, while the atmosphere indicates but 16° above zero. But, besides these, we have arguments growing out of the received theories of the distribution of temperatures upon the surface of the earth.

The actual distribution of heat in this shut-out region can only be inferred. The system of isothermals, projected by

Humboldt upon positive data, ceased at 32° ; and the views of Sir John Leslie, that the north pole was not the coldest part in the Arctic regions, have since been disproved.

Sir David Brewster, by a combination of the observations of Scoresby, Gieseke, and Parry, determined the existence of two poles of cold, one for either hemisphere, and both holding a fixed relation to the magnetic poles. These two seats of maximum cold are situated respectively in Asia and America, in longitudes 100° west and 95° east, and *on the parallel* of 80°. They differ about five degrees in their mean annual temperature ; the American, which is the lower, giving three degrees and a half below zero. The isothermals surround these points, in a system of curves yet to be confirmed by observation ; but the inference which I present to you without comment is that, between the American and Siberian centres of intensity, the climate must be milder, or, more properly speaking, the mean annual temperature must be more elevated.

Petermann, taking as a basis the data of Professor Dove, deduces a movable pole of cold, which in January is found in a line from Melville Island to the River Lena, and, gradually advancing with the season into the Atlantic Ocean, recedes with the fall of winter to its former position. Such a movement is clearly referable to the summer land currents with their freight of polar ice.

With the consolidation of winter, the ice recedes, and the Gulf Stream enters more perceptibly into the far north. The mean temperature of the northeast coast of Siberia is forty or fifty degrees colder than that of the western shores of Nova Zembla, while in July it is twenty degrees higher.

But if any point between 75° and 80° north latitude, a range sufficiently wide to include all the theories, be regarded as the seat of the greatest intensity of cold, we may, perhaps, infer the state of the Polar Sea from the known temperatures of other regions, equally distant with it from this supposed centre ; though, as the lines of latitude do not correspond with those temperature, this must be done with caution.

I have been interested for some time in examining this class of deflections, and I find that they point to some interesting conclusions as to the fluidity of the region about the pole and its attendant mildness of weather.

Thus, for instance, at Cherie Island, surrounded by moving waters, but in a higher latitude than Melville Island, the seat of the greatest observed mean annual cold, the temperature was found so mild throughout the entire Arctic winter, that rain fell there upon Christmas Day.

Barentz, a most honest and reliable authority, speaks of the increasing warmth as he left the land to the north of 77°. The whalers, north of Spitzbergen, confirm the saying of the early Dutch, that the "Fisherman's Bight" is as pleasant as the Sea of Amsterdam.

In West Lapland, as high as 70°, barley has been, and I believe is still grown ; though here is its highest northern limit. If 80° be our centre of maximum cold, the pole, at 90°, is at the same distance from it as this West Lapland limit of the growth of barley.

But there are other arguments based upon known facts, and facts, popularly recognized, bearing upon the theory of an open sea.

THE MIGRATION OF ANIMAL LIFE. - At the utmost limits of northern travel attained by man, hordes of animals of various kinds have been observed to be travelling still further.

The Arctic zone, though not rich in species, is teeming with individual life, and is the home of some of the most numerous families known to naturalists. Among birds, the swimmers drawing their subsistence from open water, are predominant ; the great families of ducks and procellarine birds throng the seas and passages of the far north, and even incubate in the regions of unknown northerness. The eider duck has been traced to breeding grounds as high 78° in Baffin's Bay, and, in conjunction with the brent goose, seen by us in Wellington Channel, and the little auk, pass, in great flights, to the northern waters beyond. The mammals of the sea, represented by the whales, and norwhal, and the seal, as well as that strange marine pachyderm, the tusky walrus, all pass in schools toward the northern waters. I have seen the white whale passing up Wellington Channel to the north, for nearly four successive days, and that, too, while all around us was a sea of broken ice.

So with the quadrupeds of these regions. The equatorial range of the polar bear is misconceived by our geographical zoologists. It is further to the north than we have yet reached ; and this powerful beast informs us of the character of the accompanying life, on which he preys. The ruminating animals, whose food must be a vegetation, obey the same impulse or instinct of far northern travel. The reindeer, although proved by my friend, Lieutenant McClintock, to winter sometimes in the Parry Group, outside of the zone of woods, comes down from the north in herds, as startling as those described by the Siberian travelers, a "moving forest of antlers."

The whalers of North Baffin's Bay, as high as 75°, shoot them in numbers ; and the Esquimaux of Whale Sound, 77°, are clothed with their furs. Five thousand skins are sent to Denmark from Egedesminde and Holsteinberg alone.

Before passing from this branch of my subject, I must mention, also, that the polar drift-ice come first from the north. The breaking up, the thaw of the ice-plain, does not commence in our so-called warmer south, but in regions to the north of those yet attained. Wrangell speaks of this on the Asiatic seas ; Parry, above Spitzbergen, and my friend, Captain Penny, confirms it in his experience of Wellington Sound.

In addition to all this, we have the observations of actual travel ; although this, confirmatory as it is, must be received with caution. Barentz saw an opening water beyond the northernmost point of Europe ; Anjou, the same beyond the Siberian Bear Islands ; and Wrangell, in a sledge journey from the north of the Kolyma, speaks of a "vast illimitable ocean," illimitable to mortal vision.

To penetrate this icy annulus, to make the "northwest passage" the northeast passage to reach the pole, have been favored dreams since the early days of ocean navigation. Yet up to this moment, complete failure has attended every attempt. One voyager, William Scoresby, passed beyond the latitude of 81° 30'. But after discarding the apocryphal voyages of the early Dutch, whose imperfect nautical observations rendered entirely unreliable their assertions of latitudes, we have the names of but two who may be said to have attained the parallel of 82° : Heinrich Hudson in 1607, and Edward Parry in our own times.

This latter navigator felt that the sea, ice-clogged with its floating masses, was not the element for successful travel, and with a daring unequaled, I think, in the history of personal enterprise, determined to cross the ice on sledges. The spot he selected was north of Spitzbergen, a group of rocks called the Seven Islands, the most northern known land upon our globe. With indomitable resolution he gained within four hundred and thirty-five miles of this mysterious goal, and then, unable to stem the rapid drift to the southward, was forced to return.

But the question of access to the Arctic pole is now brought again before us, not as in the days of Hudson and Scoresby, and Parry, a curious problem for scientific inquiry, but as an object of claiming philanthropic effort, - the rescue of Sir John Franklin and his followers. The recent discoveries by the united squadrons of De Haven and Penny, of Franklin's first winter quarters at the mouth of Wellington Channel, aided by the complete proofs since obtained that he did not proceed to the east or west, render it, beyond conjecture, certain that he passed Wellington Channel to the north. Here we have lost him ; and save the lonely records upon the tombstones of his dead, for seven years he has been lost to the world. To assign his exact position is impossible ; we only know that he has travelled up this land-locked channel, seeking the objectives of his enterprise to the north and west. That some of his party are yet in existence this is not the place to argue. Let the question rest upon the opinions of those, who having visited this region, are, at least, better qualified to judge of its resources than those who have formed their opinions by the fireside.

The journeys of Penny, Goodsir, Manson, and Southerland have shown this tract to be a tortuous estuary, a highway for the polar ice-drift, and interspersed with islands as high as latitude 77° ; beyond which they could not see. It is up this channel that the searching squadron of Sir Edward Belcher has now disappeared, followed by the anxious wishes of those who look to it as the final hope of rescue. I regret to say, that after considering carefully, the prospects of this squadron, I have to confess that I am far from sanguine as to its success. It must be remembered that Wellington Channel is all that has just been stated, - tortuous, and a thoroughfare for the northern ice ; and the open water sighted by Captain Penny is not to be relied on, either as extending very far, or as more than temporarily unobstructed. If we look up from the highlands of Beechy Head, fifty miles of apparently open navigation is all that we can assert certainly, to have been attained by the searching vessels, and to reach the present known limits of the sound would require progress, in a direct line, on their part, of at least one hundred and thirty miles.

They left, moreover, on the fifth of August ; and early as this is there considered, and open as was the season, they have but forty days before winter cements the sea, or renders navigation impossible by clogging the running gear. By a fortunate concurrence of circumstances, the squadron of Sir Edward Belcher may do everything ; but as I must repeat, that I am far from sanguine as to their success. The chances are against their reaching the open sea. It is to announce, then, another plan of search that I am now before you ; and as the access to the open sea forms its characteristic feature, I have given you the preceding outline of the physical characteristics of the region, in order to enable you to weigh properly its merits and demerits.

It is in recognition of the important office which American geographers may perform toward promoting its utility and success, that I have made the Society the first recipient of my plan.

My plan of search is based upon the probable extension of the land masses of Greenland to the far north, - a view yet to be verified by travel, but sustained by the analogies of physical geography. Greenland is a peninsula, and follows, in its formation, the general laws which have been recognized since the days of Forster as belonging to peninsulas with a southern trend. Its abrupt, truncated termination at Staaten Hook, is as marked as that which is found at the Capes Good Hope and Horn of the two great continents, the Comorin of Peninsular India, Cape South East of Australia, or the Gibraltar of Southern Spain.

Greenland is lined by a couple of lateral ranges, metamorphic in structure, and expanding in a double axis to the NNW and

NNE. They present striking resemblances to the Ghauts of India, being broken by the same great injections of greenstone, and walling in a plateau region where glacial accumulations correspond to those of the Hindostan plains.

The culmination of these peaks in series, indicates strongly their extension to a region far to the north. The same continued elevation is observed by the whalers as high as 77°, and Scoresby noted nearly corresponding elevations on the eastern coast, in latitude 73°. From these alternating altitudes, continued throughout a meridian line of nearly eleven hundred geographical miles, I infer that Greenland is continued further to the north than other known lands. Believing, then, in such an extension of Greenland, and feeling that the search for Sir John Franklin is best promoted by a course which will lead directly to the open sea, will be most likely to afford some trace of the lost party, I am led to propose and attempt this line of search.

Admitting such an extension of the land masses of Greenland to the north, we have the following inducements for exploration and research : -

1. Terra firma as the basis of our operations, obviating the capricious character of ice travel.

2. A due northern line, which would lead soonest to the open sea, should such exist.

3. The benefit of the fan-like abutment of land, on the north face of Greenland, to check the ice in course of its southern or equatorial drift, thus obviating the great drawback of Parry in his attempts to reach the pole by Spitzbergen Sea.

4. Animal life to sustain travelling parties.

5. The cooperation of the Esquimaux.

The point I would endeavor to attain would be the highest attainable seats of Baffin's Bay, from Smith's Sound, and advocated by Wrangell as the most eligible site for reaching the pole.

As a point of departure it is two hundred and twenty miles ti the north of Beechy Island, the starting-point of Sir Edward Belcher, and seventy miles north of the limits seen or recorded in Wellington Channel.

We shall quit the United States in time to reach the bay at the earliest season of navigation. The brig furnished by Mr. Grinnell for this purpose is admirably strengthened and fully equipped to meet the peculiar trials of the service. After reaching the settlement of Upernavik, we take in a supply of Esquimau dogs, and a few picked men to take charge of the sledges.

We then enter the ice of Melville Bay, and if successful in penetrating it, hasten to Smith's Sound and secure our vessel for the winter. The operations of search, however, are not to be abandoned. I am convinced we can push forward our provision depots by sledge and launch, and thus prepare for the final efforts of our search.

In this I am strengthened by the valuable opinion of my friend, Mr. Murdaugh, late the sailing-master of the *Advance*. He has advocated this very sound as a basis of land operation. And the recent journey of Mr. William Kennedy, commanding Lady Franklin's last expedition, shows that the fall and winter should no longer be regarded as lost months.

It is my intention to cover each sledge with a gutta percha boat, a contrivance which the experience of the English has shown to be perfectly portable. Thus equipped, we follow the trend of the coast, seeking the open sea.

Once there, if such a reward awaits us, we launch our little boats, and, bidding God speed us, embark upon its waters.

VEGETATION AND THE INSIDE OCEAN

According to Schwartz and Liuk and acre of meadow-land where it is wet, produces 4,400 pounds of hay, which, when dried, contains forty-six percent of carbon. The hay, then, yields 2,000 pounds of carbon, to which 1,000 pounds should be added, for the portion of the season in which the grass is not cut, and also for the roots. To produce these 3,000 pounds of carbon, 12,000 pounds of carbonic acid is required. Schubler has demonstrated that an acre of grass, as poor a kind as Poa annua, exhales in one hundred and twenty days of active vegetation 6,000,000 pounds of water. To supply this demand of carbon it is necessary for the ground to imbibe three and a half grains of carbonic acid with every pound of water. Mr. Lawes has found that in a plant of any ordinary crop, more than two hundred grains of water must pass through it, for a single grain of solid substance to accumulate within it.

Nitrogen comes next to be considered. The yield of this element to vegetation seems to be independent of manures. A water meadow, it has been estimated, which has never received any manure, yields annually about forty-five pounds of nitrogen, while the best ploughed lands yields only about thirty pounds. That there is a supply independent of the soil, is seen in the great quantities of nitrogenous matters - hay, butter, and cheese - which are carried off from the lands without any diminution of the supply of nitrogen. The estimate of ammonia is one-thirteenth of a grain in every pound of water for the exigencies of vegetation.

No spring-water contains so small an amount as this estimate. The great desideratum is to bring the soil into harmony with the conditions by which the nutrition of plants may best be promoted. Much depends upon the nature of the soil. Manure produces or yields a small amount of inorganic matter to the soil and gives it a higher degree of temperature ; these are the principal benefits of manure to vegetation ; but as to nutrition, manure yields but little to the growth of plants. The darkest colored lands are generally the highest in temperature ; hence the advantage of all vegetable moulds. Deep, light sands and clay, which turn almost to stone in dry weather, are unproductive. The application of humus evolves heat by the process of combustion. The combinations needed are lime, clay, and humus ; the clay being in proportion of forty-five percent ; if less than ten percent, the land will be light and poor. It is only necessary that the soil be ploughed as far as necessity requires. By too frequently loosening of the soil, the decomposition of humus is so rapid as to overbalance the benefit derived from the exposure to the atmosphere. This leads to the discussion of fallowing. There are two kinds of fallow, - naked and covered. As a general rule, covered fallows are preferable to naked fallows, as the latter tend to waste the nutritive elements in the soil required for vegetation. In the covered fallows, that is those sown with clover, the quantity of humus and carbonic acid is increased by the clover, preventing evaporation of the nutritive elements from the soil. Naked fallowing is only to be resorted to when there is no other way of loosening the soil. In an acre of clover during its growth, over 114,860 gallons of water are evaporated ; but this does not exhaust the soil, as two hundred grains of water must pass through the vegetation to retain one grain of solid matter, beside having left the soil in prepared condition to grow a crop. Covered fallows are preferable to manures, for the manures do not act immediately on vegetation by means of their organic constituents, but by reason of their warmth and of the inorganic substances which they involve.

M. Baudrimont states that there is a natural process at work, by which liquid currents rise to the surface from a certain depth in the ground, and thus bring up materials that either help to maintain its fertility or to modify its character. This theory will account for the improvements which take place in fallows ; and there is reason to believe that this natural process, as he terms it, materially influences the rotation of crops. Then we are led to conclude that the vegetable nutrition is mainly derived from water and the air, which are conveyors of oxygen, nitrogen, hydrogen, and carbon, the four elements of plants.

The author's views are that vegetation when growing receives from 90 to 92 percent of water, with the remainder, lime 3 parts, potash 2 3/4, ammonia 30, in its natural state. Wheat 34 percent of solid substance, and other green substances from 30 to 35 percent, when matured. If we admit these statements, and others too numerous to mention, where shall we find water sufficient to supply this vast quantity but in the principle of the water rising from the ocean inside where rising through pores in the earth to the air, and striking the roots of vegetables, and ascending to the topmost leaf, kept from the air by the bark ; and the water this rising through the earth connecting with every mineral and forcing them up to feed every plant and giving life to everything that exists.

CLOSING AND CONCLUSION

There are other evidences too numerous to be brought into this small work. I will close with a few remarks in general to the northern and southern extremities of the earth, which we consider as 70° the rounding point, which would leave a hole at each pole of 3,000 miles in diameter, sufficient to let air, rays of the sun, and space, pass through the centre, evidence showing the extreme. 70° is the coldest point found, also from 70° north to 70° south we get the revolution of the tide ; also Morton and Haynes, when at 81°, or nearly that, found an open ocean, the tides from the inside setting south, also currents of wind blowing two days strong from the north, which were bare the 20th of June ; while all south were covered with ice and snow ; now if they had no snow on the mountains they must have fertile valleys. So I draw this work to a close by wishing a candid investigation of it.

EXTRACTS FROM LETTERS RECEIVED BY THE AUTHOR:

Palace of the Tuileries, Paris, France

Professor Merrill, Flume House, New Hampshire, USA:

Having received from our Royal Cousin, Victoria of England, a schedule of your valuable discovery in relation to the present position of Sir John Franklin, in connection with your learned theory in relation to the condition of the interior of the earth, thereby refuting entirely the absurd notion that now prevails that McClintock's expedition made any discovery in relation to his fate, I desire to inform you that by my orders the Scientific Board of France are thoroughly investigating the fact, and request to know if, by our sending a war steamer for your honorable self, you could be induced to leave your native land for a time and allow the light of your resplendent genius to aid the united learning of France and England, at a meeting of savants to be held at our palace at Versailles the coming summer.

You will please address us at length immediately on this subject.

Given under our royal seal at the palace of the Tuileries, this 20th of March, 1860.

Louis Napoleon

Palace of the Tuileries, Paris, France, 1862

The Emperor having received from your honor pleasing intelligence and valuable documents, desires through me to express his most heartfelt thanks for your kindness, and is sensible of the great benefit you are destined to perform both in the old and new world, and that your project meets the entire approbation of all the scientific men to whom it has be subjected.

Louis Napoleon

Royal Dispatch from Her Majesty, 1857

Victoria Regina and Albertis Princess

To his August Highness Honorable John Merrill, Director of the Pool, Arctic Philosopher, and Practical Philanthropist, etc., etc.:

Monsieur, I am commanded by Her Gracious Mejesty's Highness, to communicate to your obsequious Highness the most trans-Atlantic compliment of Adel Kada, and to acknowledge the receipt of your most learned alliloquent, and circumambient State Document, dated August 28th, 1854, which has been under the consideration of the Grand Lama ever since. The Grand Lama fully concurs in your new views of the hole in the earth. The Grand Butler takes this opportunity to express to your obsequious Highness the great satisfaction which the most Grand Lama feels after perusal of so learned a document, and begs to salute you as a man of transcendent prognostications. By Royal command my own Royal pleasure. Signed in the Grand Culinary Department with Royal Goosequill.

Victoria by Albert

Tongo, China, September 22nd, 1862

My Dear Sir:

The High Mendaris of China have learned of your great knowledge and immense powers of perception, surpassing all other foreigners, and the equal in the divine wisdom of our own teachers. We are pleased to know that you have, after years of intense application, assured the learned, as well as yourself, of the existence, beyond a doubt, of an open Polar Sea, and that the lamented Franklin is still alive and within the sunless depths. From lucid explanation by you, given to his special messenger in chief, Yang tisang.

The important theory will be communicated to Yang tisang, who promises to forward to you some souvenirs of his Royal esteem, and permits you to address him at length, that he may communicate it to his millions of men. You are permitted to make use of his Royal messenger, and some line from you accepting his preferred aid and sympathy will be thankfully received and transmitted to him by the writer. Br profoundly happy and wise, man of the mountain, for the great Mogang, Yang tisang, has said it.

Written by Tafing Wang, Minister from China

Your theory is most profoundly wise.

Prince Saxe Coberge

We came to see you on your great theory of the earth.

Emma Regina Carlotta, wife of Maximilian

... and others too numerous to mention.

This republish of Cosmogony is dedicated to
John Merrill, the Philosopher of the Pool ...

... the path less travelled ...

... and the Merrill family, who are often found there.